How to use this book

AF604102

A sample page

INSTRUCTION
What your child needs to do for the activity.

TITLE
The page title describes the skill your child will learn in these pages.

FUN ILLUSTRATIONS
Specially drawn illustrations which are fun, interesting and drawn at the right pedagogical level for your child.

COLOURFUL BORDERS
The page borders make each page as attractive as possible to stimulate your child.

EXAMPLE
The first one is done for you so you can show your child exactly what to do.

LOTS OF PRACTICE
Two pages where your child can practise and repeat the same skill to master it.

STICKERS
Place a sticker on each page as your child finishes.

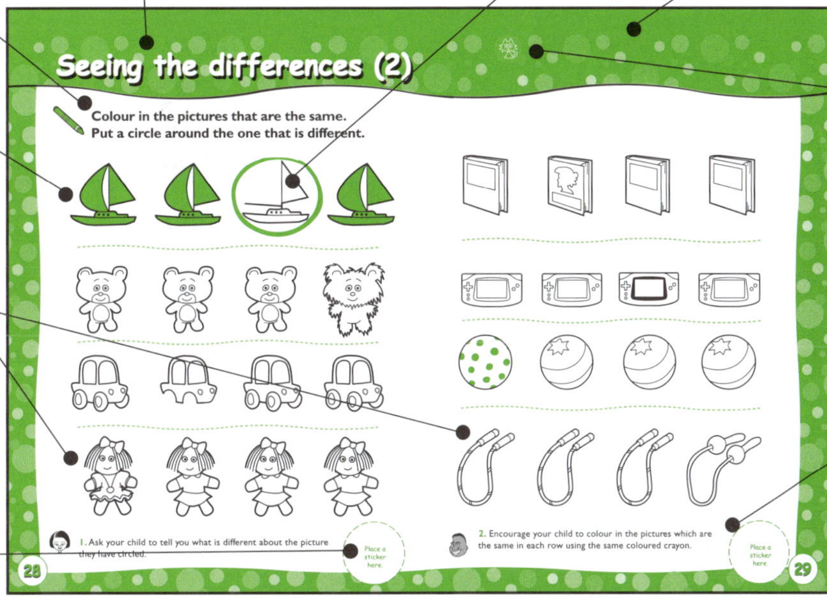

WHO'S HIDING?
In each book, a little creature appears in the border of every double page so your child can have fun trying to find it.

EXTRA ACTIVITIES
Extra activities you might want to do with your child to further reinforce the skill or simply make it more enjoyable.

Step-by-step learning

STEP ONE **Read** out the title of the activity page to your child.

STEP TWO **Explain** the skill and show your child the example already done. **Make sure** they understand what to do. Your child will then have at least two pages to practise that same skill.

STEP THREE **Help** your child put a **sticker** on the bottom of each page as they complete it.

Remember to be patient, encouraging and positive with your child, even when minor mistakes are made!

How to hold a pencil

It is important that you help your child hold his or her crayon or pencil in the correct way as shown here to ensure your child develops the right technique early on.

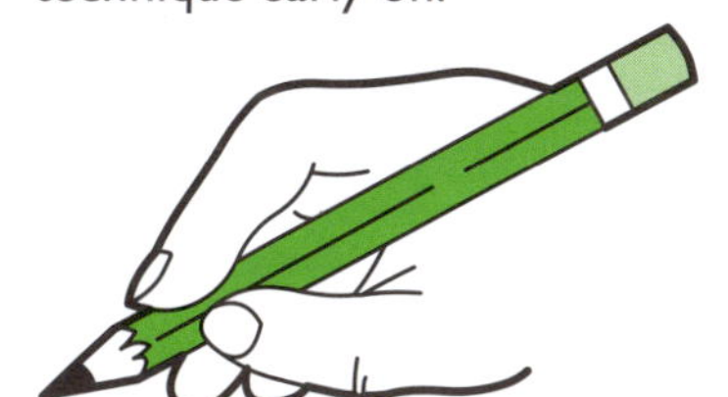

Drawing stripes

Trace over the dotted lines to make a pattern of stripes.

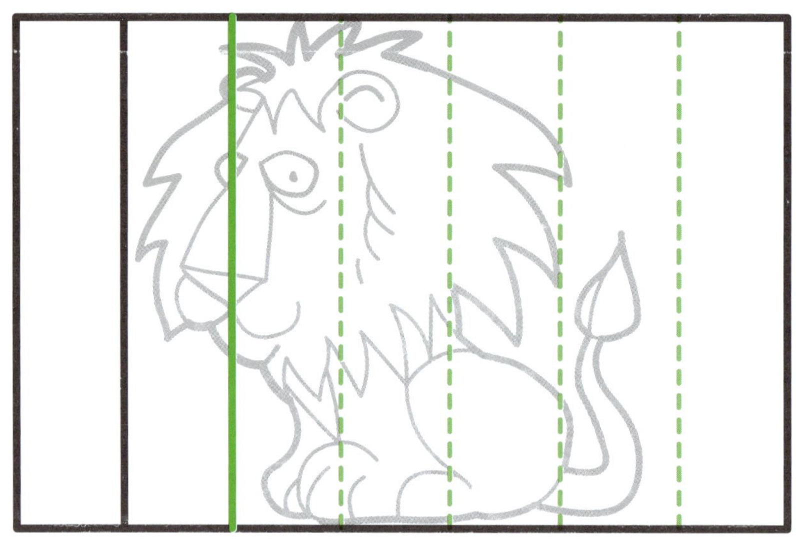

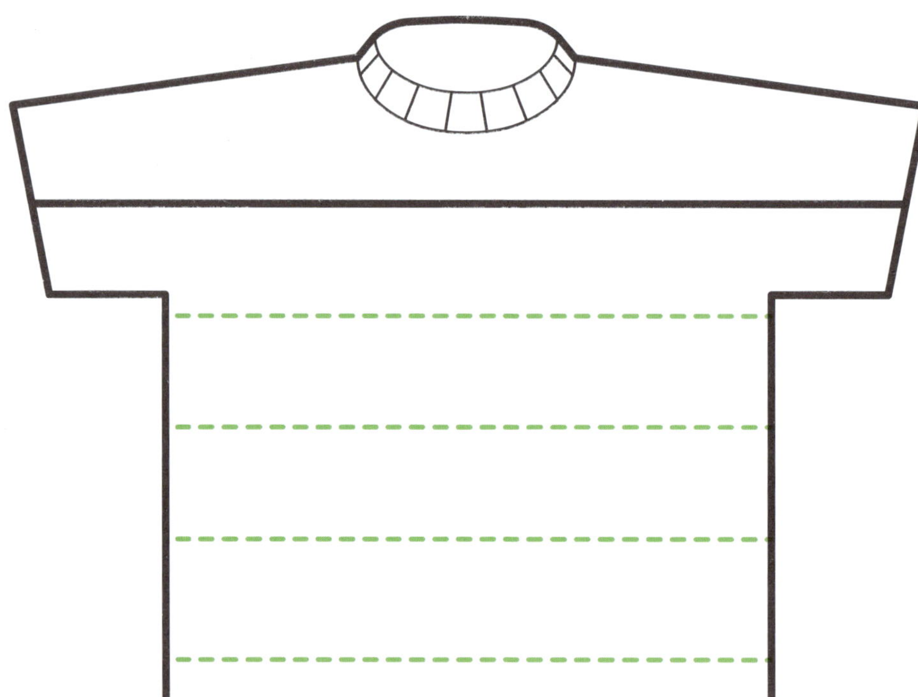

1. Ask your child to look for other striped objects.

Place a sticker here.

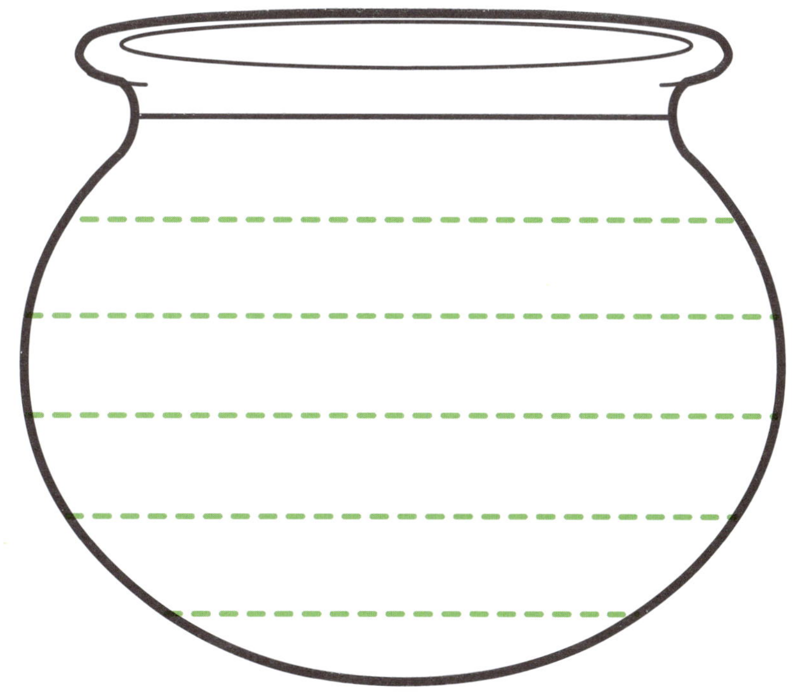

2. Find any of your child's clothes which have stripes on them and have them trace over the stripes with their fingers.

Place a sticker here.

Drawing circles

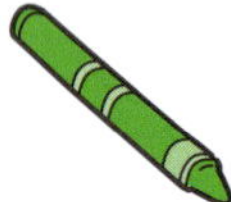

Trace over the dotted lines to make a picture or pattern of circles.

1. Where else can you see circles around you? Ask your child to find some.

Place a sticker here.

2. Ask your child to count the circles on each page.

Place a sticker here.

Copying patterns (1)

Trace over each pattern, going all the way across the two pages.

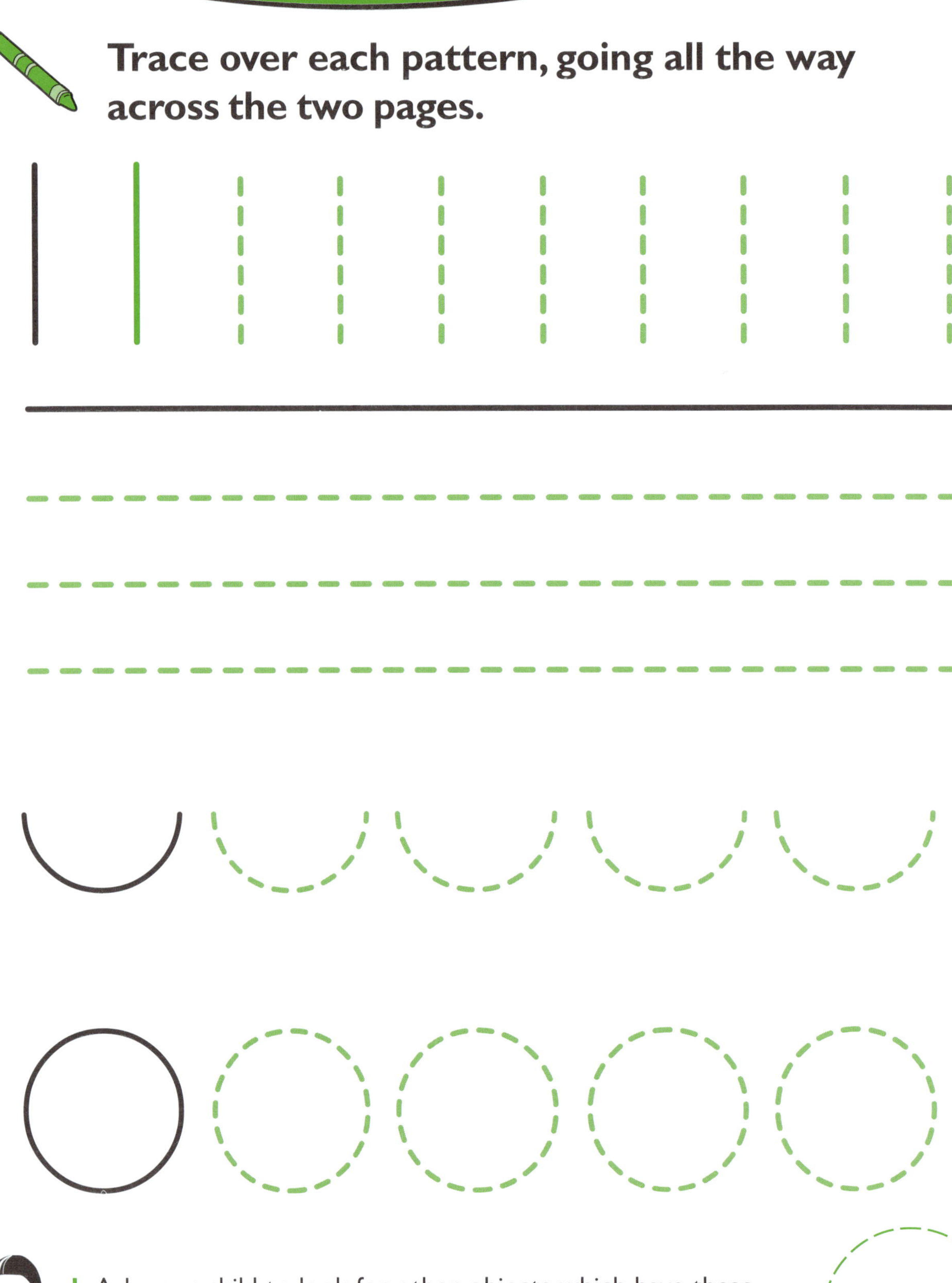

1. Ask your child to look for other objects which have these patterns on them.

Place a sticker here.

2. Have your child use a different coloured crayon or pencil for each pattern, and ask them to name each colour.

Place a sticker here.

Colouring patterns (1)

Pick two different coloured crayons. Colour the first shape in one colour, then the second shape in the next colour. Then make the same pattern again until you finish the row.

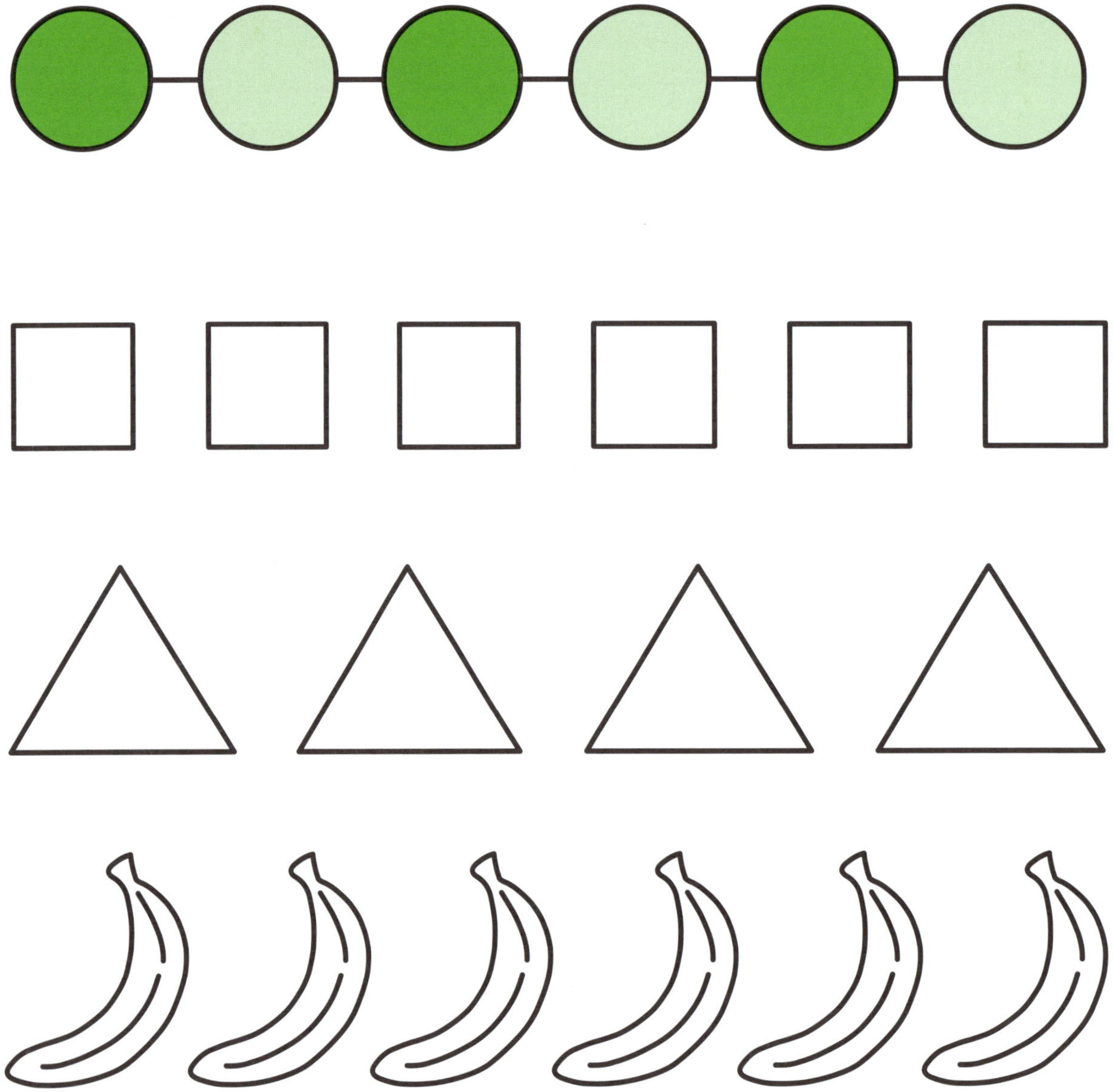

1. Try to make a shoe, sock, shoe, sock pattern on the floor.

Place a sticker here.

2. Make it a regular adventure to look for patterns around the home or when you are out walking.

Place a sticker here.

Finishing patterns (1)

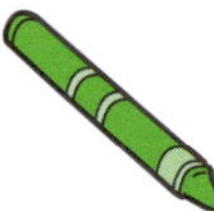

Each snake has a different pattern. Trace over the dotted lines, then make and continue each pattern yourself.

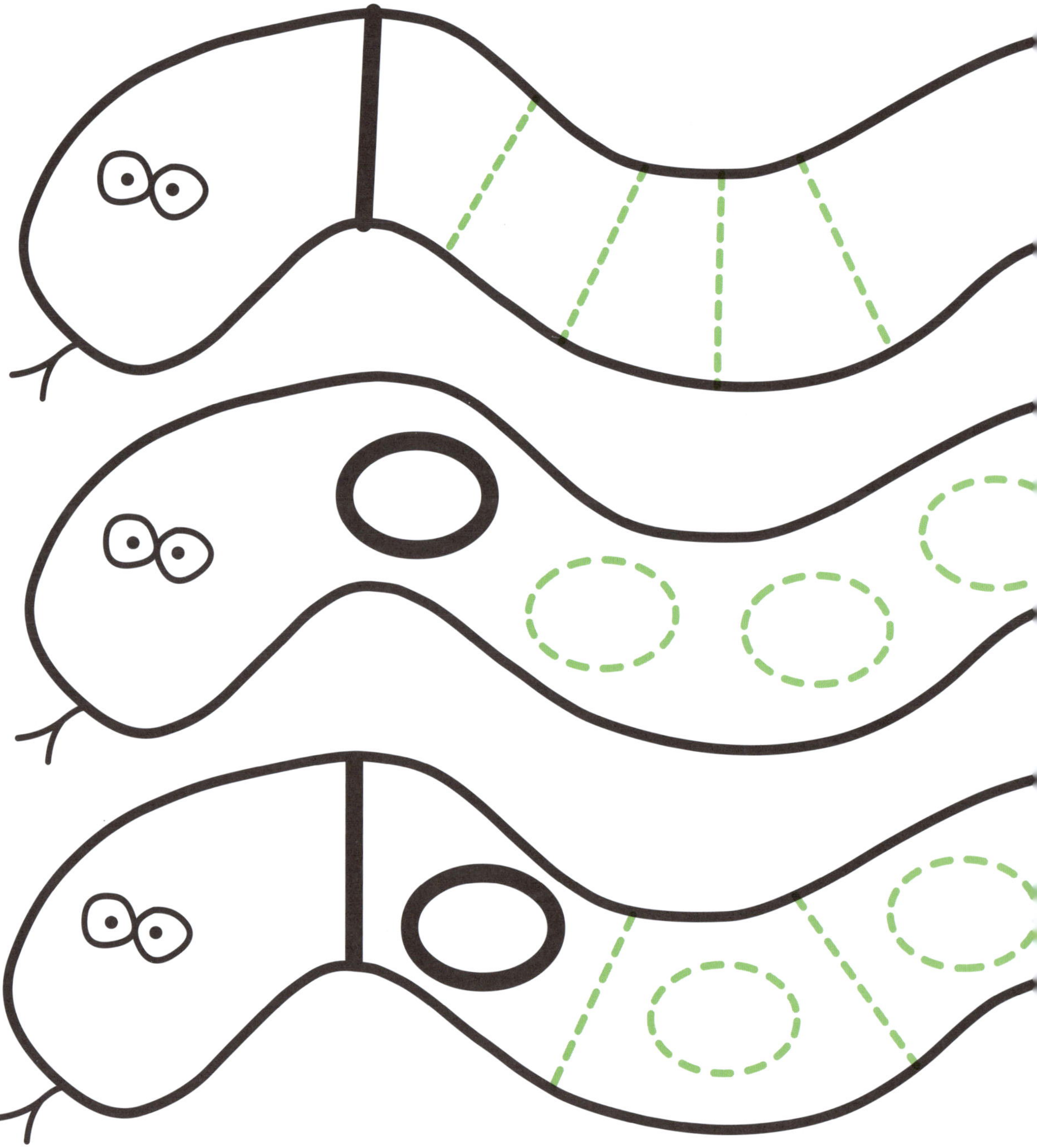

1. See if you can find pictures of snakes with other patterns on them to show your child.

Place a sticker here.

2. Ask your child to count how many snakes are on the page.

Place a sticker here.

Copying patterns (2)

Look at each pattern on the first page. Then trace over the lines to make the same pattern on the next page.

1. Ask your child to describe each pattern. What are the names of the shapes?

Place a sticker here.

2. Make a happy or sad face pattern with your face too!

Place a sticker here.

Colouring patterns (2)

Pick two different coloured crayons. Colour the first two shapes in one colour, then the second two shapes in the next colour. Then make the same pattern again until you finish the row.

1. Try to make a toy, toy, book, book repeating pattern on the floor.

Place a sticker here.

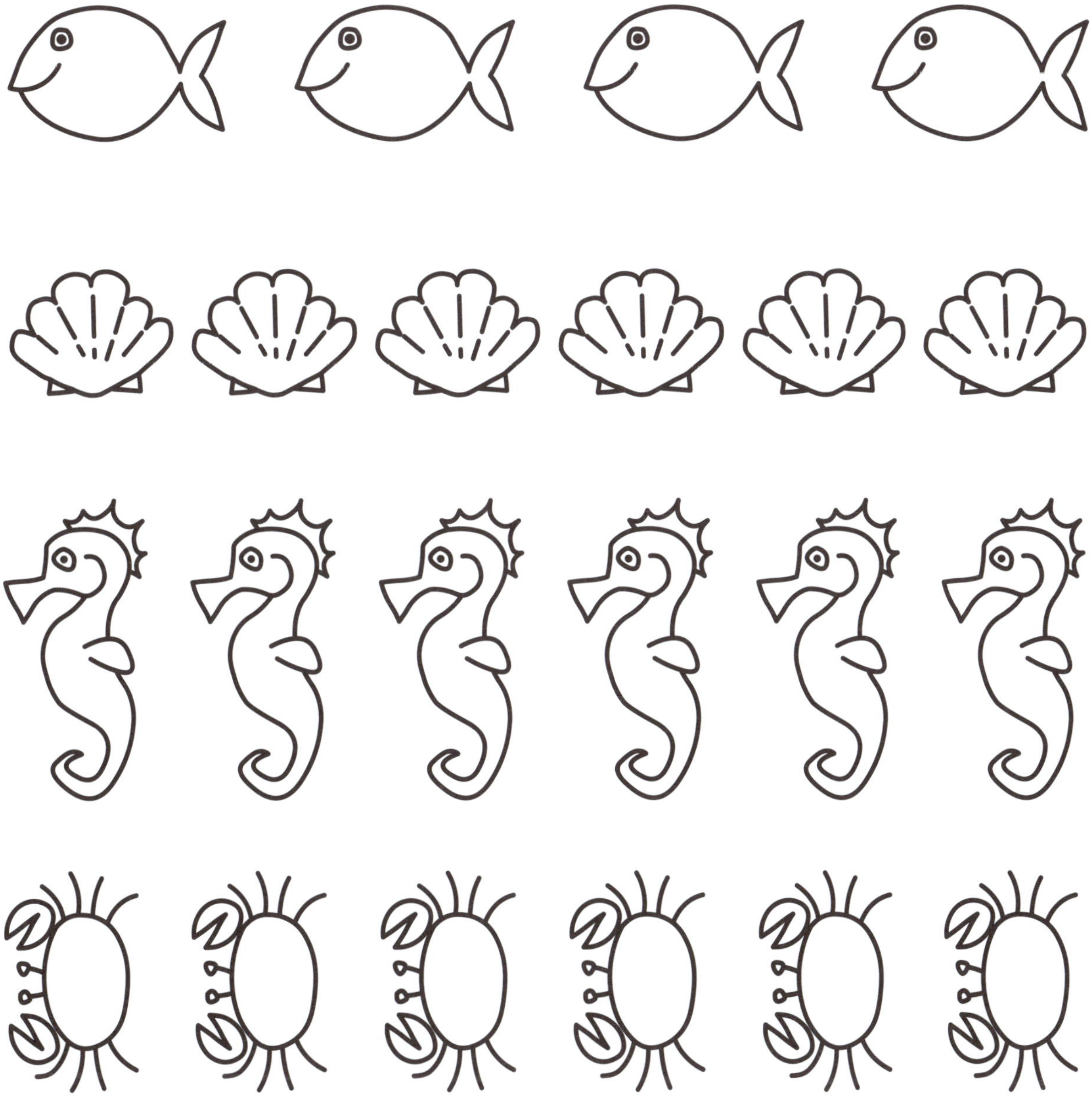

2. Ask your child to name the pictures in each row on this page. Where would all these things be found?

Place a sticker here.

Finishing patterns (2)

Each snake has a different pattern. Trace over the shapes, then make each pattern yourself.

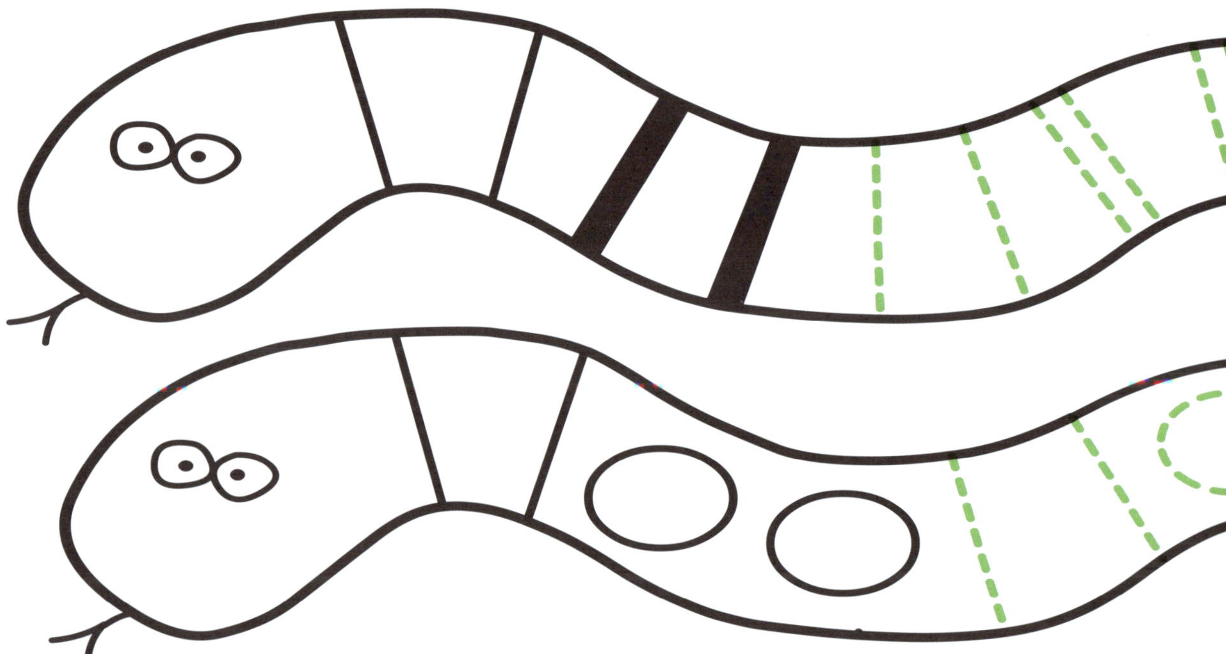

Each string of beads has a different pattern. See if you can copy the pattern and finish off the beads.

1. Can you show your child a real necklace or string of beads with a pattern on it?

Place a sticker here.

2. Your child can use a different coloured crayon for each pattern.

Place a sticker here.

Matching pictures (1)

Colour the pictures that are the same in the same colour.

1. Ask your child to name the pictures on each page as they colour them in.

Place a sticker here.

2. Ask your child to tell you where these groups of objects can be found (ie. in a farmyard, at the beach). What other things might also belong in each group?

Place a sticker here.

Matching pictures (2)

Colour the pictures that are the same in the same colour.

1. Ask your child to name the pictures on each page as they colour them in.

Place a sticker here.

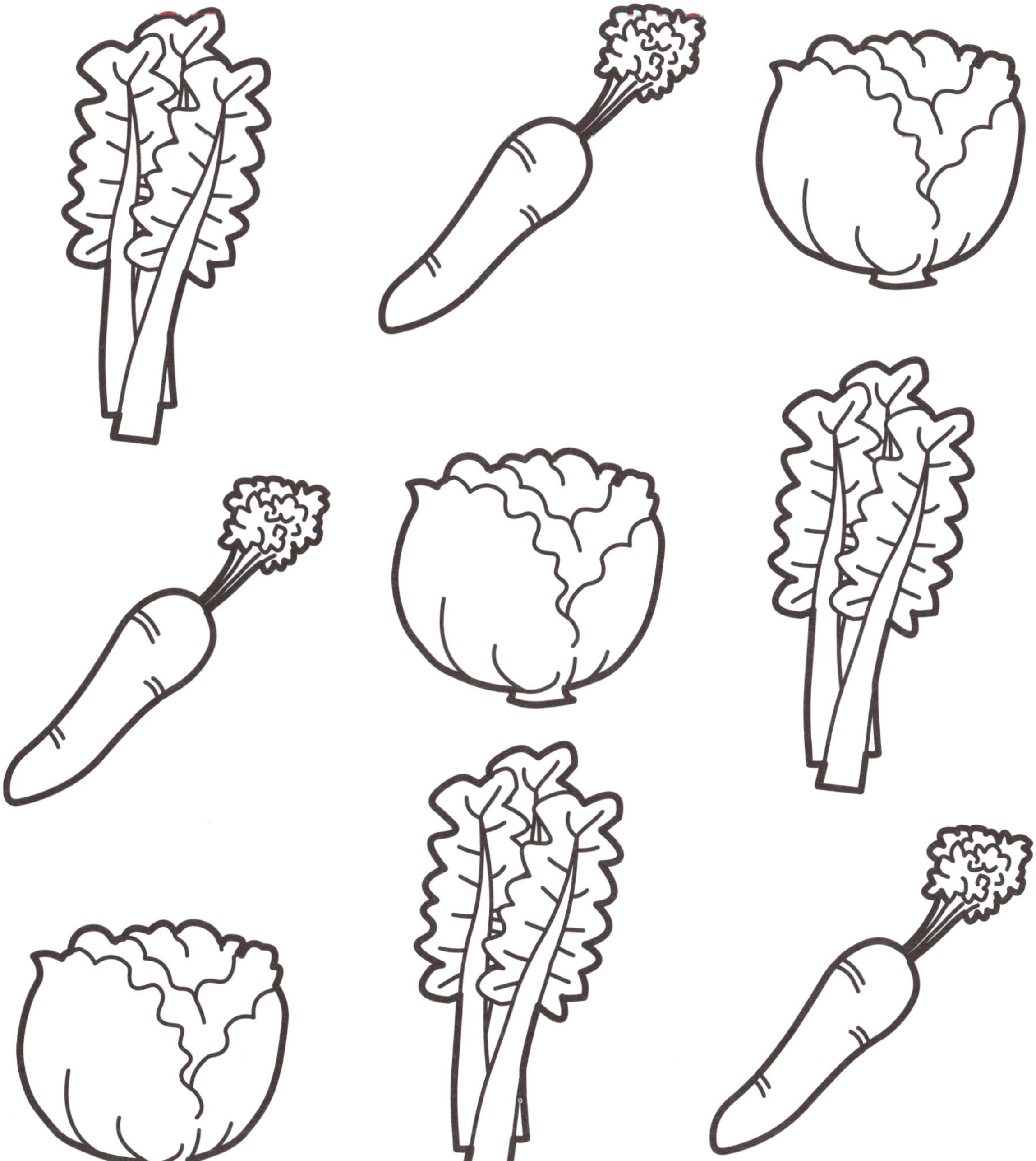

2. Ask your child to count how many of each picture is on the page.

Place a sticker here.

Matching pictures (3)

Colour the pictures that are the same in the same colour.

1. Ask your child to name the pictures on each page as they colour them in.

Place a sticker here.

2. Ask your child to tell you how many things are in each group.

Place a sticker here.

Sorting into groups

There are 3 green fish in the bowl. Colour 3 fish in red and 4 in yellow. Then count the fish in each group.

1. Show your child a group of coloured objects (eg. a set of coloured building blocks) and see if they can sort them into groups of the same colour.

Place a sticker here.

Colour 2 balloons in red, 4 in green and 4 in yellow. Then count the balloons in each group.

2. Show your child a mixed group of different coloured objects (eg. cups, blocks, small containers, toys) and see if they can sort them into groups of the same colour.

Place a sticker here.

Seeing the differences (1)

Colour in the pictures that are the same.
Put a circle around the one that is different.

1. Ask your child to tell you about the picture which is different.

Place a sticker here.

2. Look for objects which are the same or different around the home.

Place a sticker here.

Seeing the differences (2)

Colour in the pictures that are the same.
Put a circle around the one that is different.

1. Ask your child to tell you what is different about the picture they have circled.

Place a sticker here.

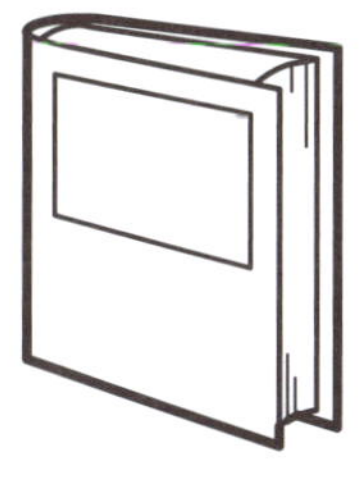

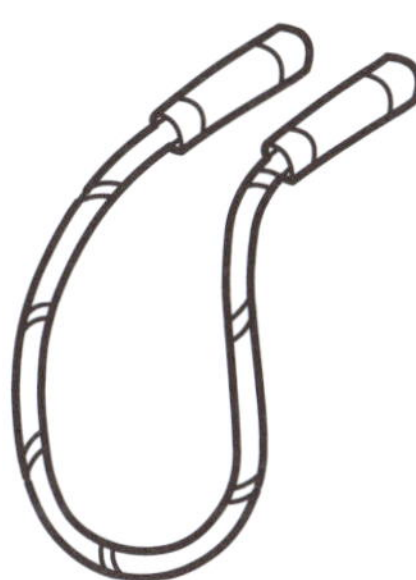
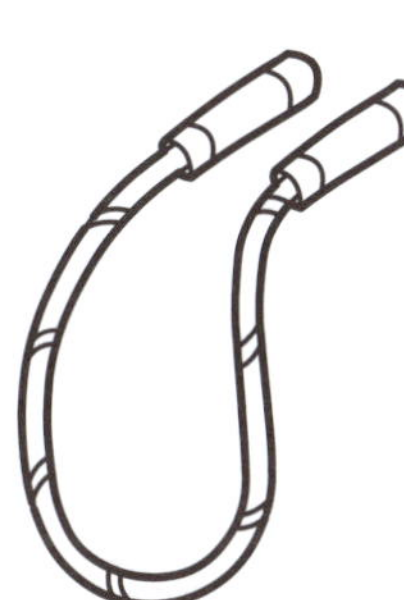
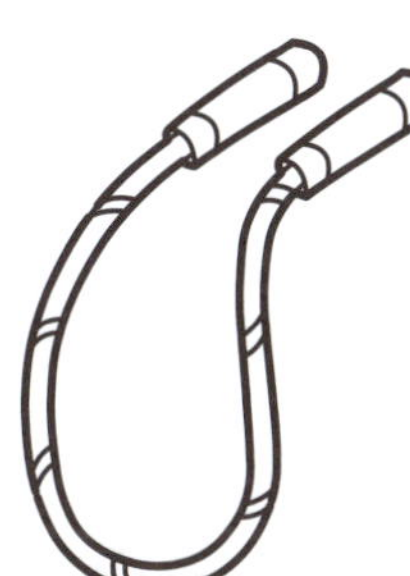
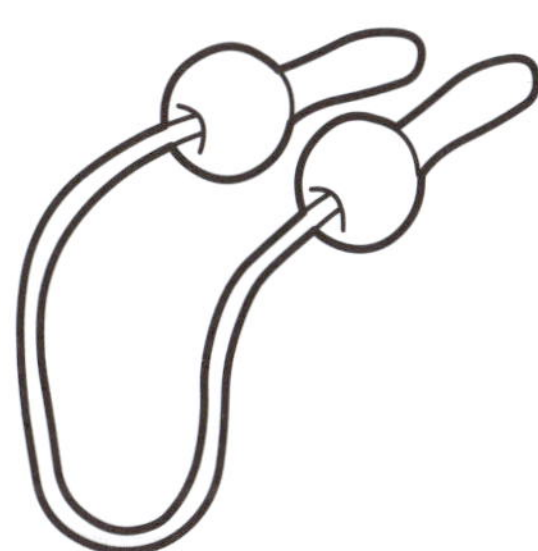

2. Encourage your child to colour in the pictures which are the same in each row using the same coloured crayon.

Place a sticker here.

Seeing the differences (3)

Look at each picture. Can you find *five* differences between the top and bottom pictures? Put a circle around each one.

1. Ask your child to tell you about each difference as he or she finds it.

Place a sticker here.

2. Your child can draw one more thing which is different in the bottom picture.

Place a sticker here.

Matching one-for-one (1)

Draw a line to join up the things that go together.

Join each calf to a cow.

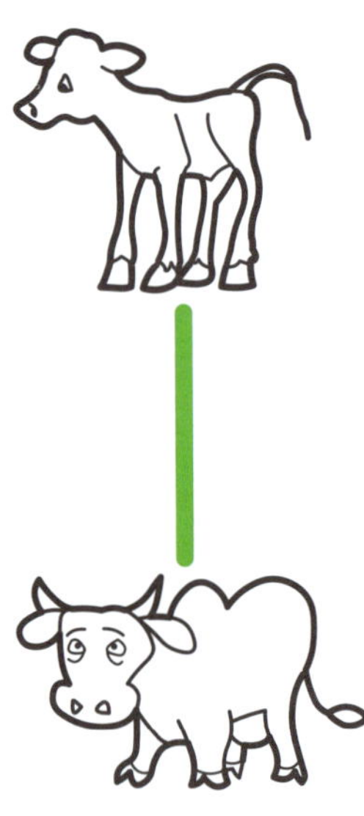

Join each girl to a boy.

1. Ask your child to name the pictures and say why they go together.

Place a sticker here.

Join each sock to a shoe.

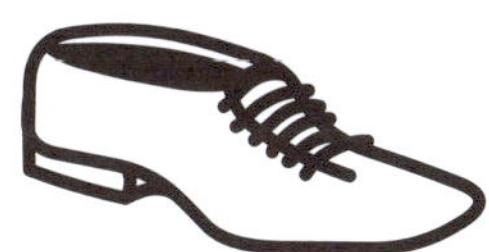

Join each hand to a glove.

2. Is there a cow for each calf? A shoe for each sock?

Place a sticker here.

Matching one-for-one (2)

Draw a line to join up the things that go together.

Join each dog to a bone.

Join each apple to a worm.

1. Ask your child to name all the pictures.

Place a sticker here.

Join each fork to a knife.

Join each cup to a saucer.

2. Ask your child if there is a knife for each fork and a saucer for each cup.

Place a sticker here.

Drawing the missing parts

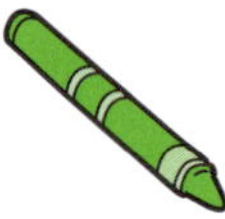

Draw the missing part for each picture.

Draw a rattle for each baby.

Draw a hat for each girl.

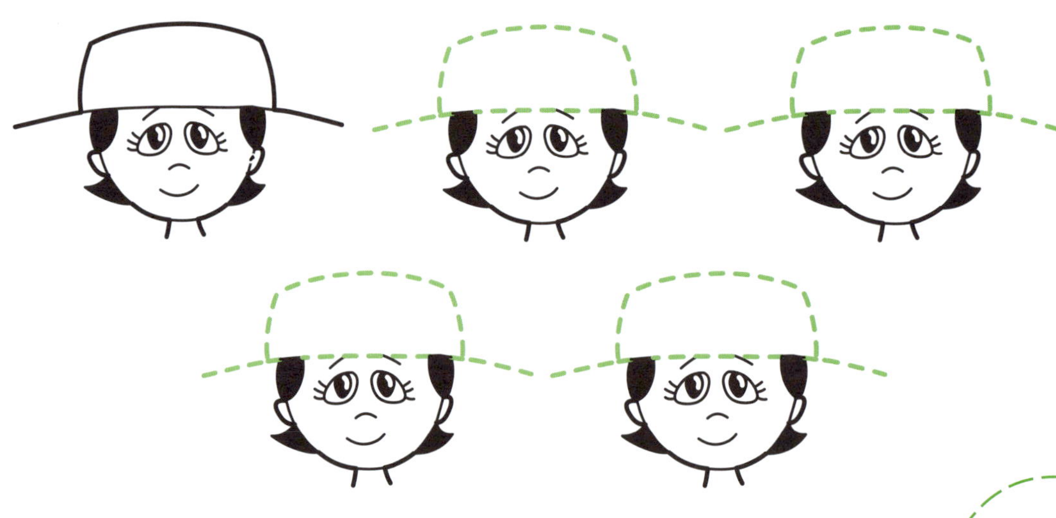

1. Ask your child to name all the pictures.

Place a sticker here.

Draw a roof for each house.

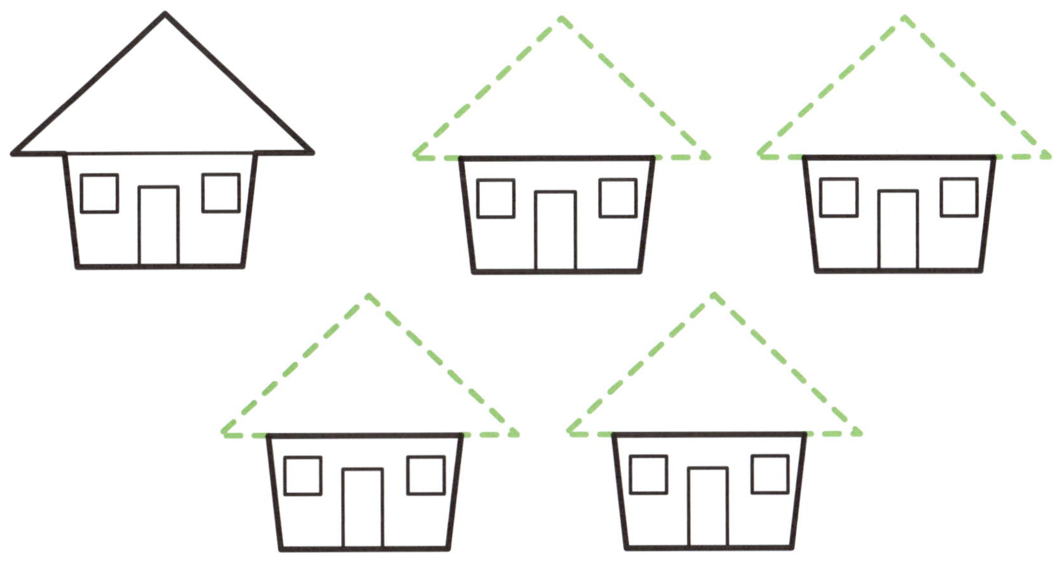

Draw a patch for each pirate.

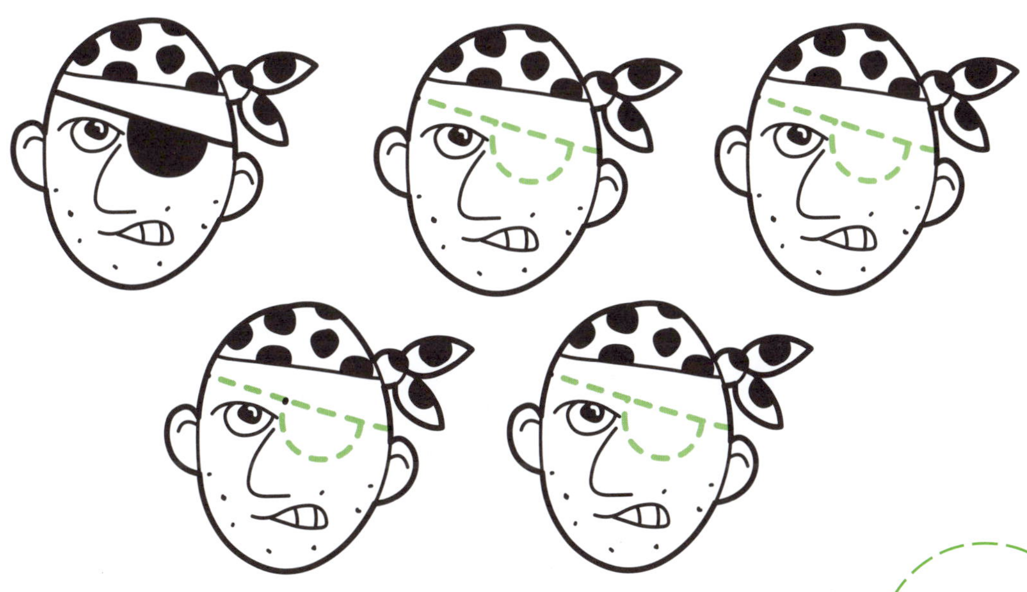

2. Ask your child if there is a rattle for each baby and a patch for each pirate.

Place a sticker here.

Drawing the missing parts

Draw the missing part for each picture.

Draw a spot for each cow.

Draw a tail for each cat.

1. Ask your child to say what part is missing on each picture.

Place a sticker here.

(2)

Draw a smile for each boy.

Draw an ice-cream for each cone.

2. Ask your child to count how many things are in each group.

Place a sticker here.

Drawing the missing parts

Draw the missing parts on the second picture in each pair.

1. Ask your child to tell you what's missing before drawing it in.

Place a sticker here.

2. Your child can colour in the pictures.

Place a sticker here.

Drawing the missing parts

Draw the missing parts on the second picture in each pair.

1. Ask your child to count how many objects they need to draw in.

Place a sticker here.

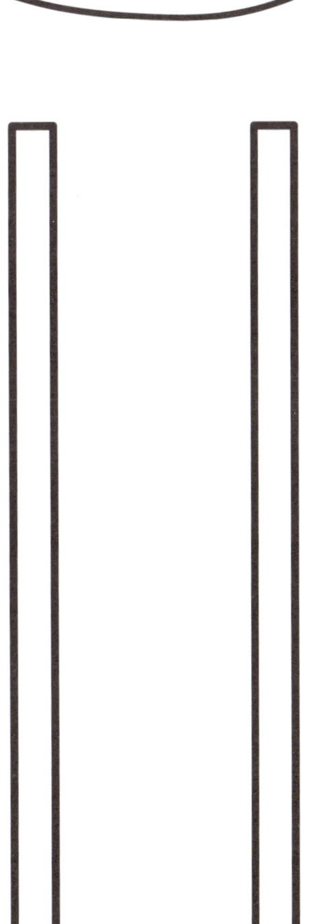

2. Ask your child what is missing from each picture before they draw it in.

Place a sticker here.

Drawing the same number

Count how many things are in each picture.
Draw in the same things in the space.

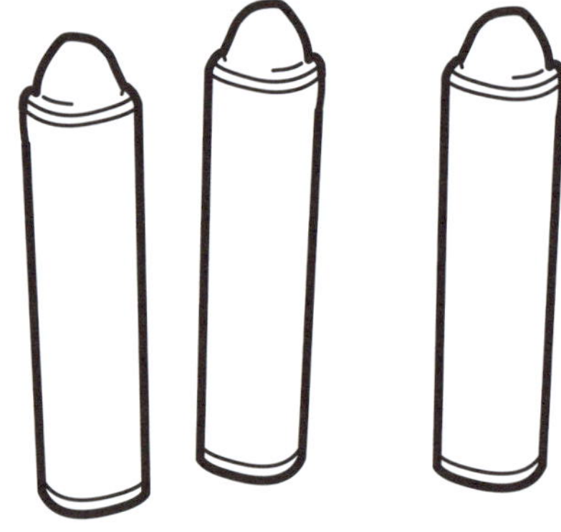

1. Ask your child to count the things out loud and tell you what they are.

Place a sticker here.

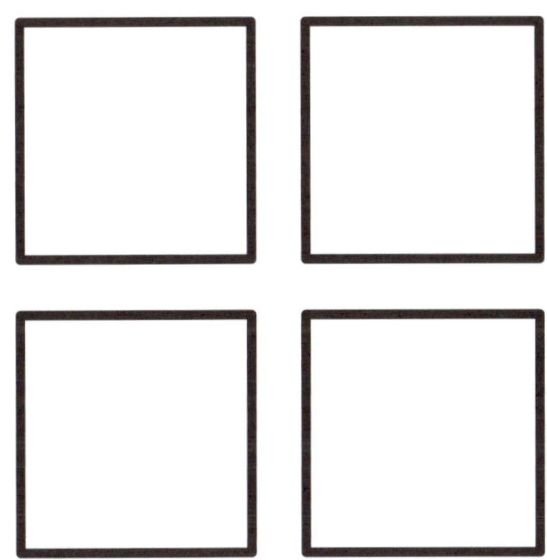

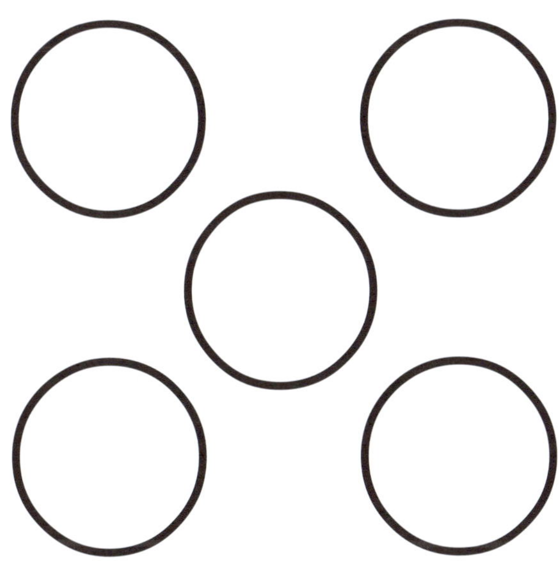

2. Ask your child what the shapes on this page are called. Can you see any of them around you in the home?

Place a sticker here.

Drawing the same number

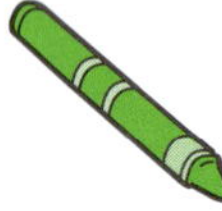

**Count how many things are in each group.
Draw in the same things in the space.**

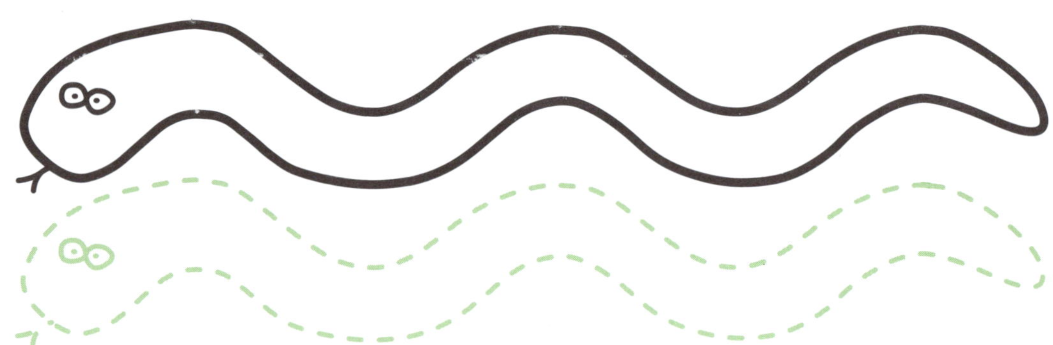

1. Ask your child to count the things out loud and tell you what they are.

Place a sticker here.

2. Ask your child to find sets of 2, 3, 4 and 5 objects from around your home.

Place a sticker here.

Well done!

You have finished the book!

Place your last two stickers on the picture.
You can colour in the picture, too.